CONTENTS

STATEMENTS

S. HRG. 104–75

FERC ELECTRIC UTILITY RESTRUCTURE

HEARING

BEFORE THE

COMMITTEE ON ENERGY AND NATURAL RESOURCES UNITED STATES SENATE

ONE HUNDRED FOURTH CONGRESS

FIRST SESSION

ON THE

FEDERAL ENERGY REGULATORY COMMISSION'S ELECTRIC UTILITY RESTRUCTURING PROPOSAL: PROMOTING WHOLESALE COMPETITION THROUGH OPEN-ACCESS NON-DISCRIMINATORY TRANSMISSION SERVICES BY PUBLIC UTILITIES (DOCKET NO. RM95–8–000) AND RECOVERY STRANDED COSTS BY PUBLIC UTILITIES AND TRANSMITTING UTILITIES (DOCKET NO. RM94–7–001)

MAY 10, 1995

Printed for the use of the
Committee on Energy and Natural Resources

U.S. GOVERNMENT PRINTING OFFICE

91–929 CC WASHINGTON : 1995

For sale by the U.S. Government Printing Office
Superintendent of Documents, Congressional Sales Office, Washington, DC 20402
ISBN 0-16-047392-6

91-929 O - 95 - 1

COMMITTEE ON ENERGY AND NATURAL RESOURCES

FRANK H. MURKOWSKI, Alaska, *Chairman*

MARK O. HATFIELD, Oregon
PETE V. DOMENICI, New Mexico
DON NICKLES, Oklahoma
LARRY E. CRAIG, Idaho
BEN NIGHTHORSE CAMPBELL, Colorado
CRAIG THOMAS, Wyoming
JON KYL, Arizona
ROD GRAMS, Minnesota
JAMES M. JEFFORDS, Vermont
CONRAD BURNS, Montana

J. BENNETT JOHNSTON, Louisiana
DALE BUMPERS, Arkansas
WENDELL H. FORD, Kentucky
BILL BRADLEY, New Jersey
JEFF BINGAMAN, New Mexico
DANIEL K. AKAKA, Hawaii
PAUL WELLSTONE, Minnesota
HOWELL HEFLIN, Alabama
BYRON L. DORGAN, North Dakota

GREGG D. RENKES, *Staff Director*
GARY G. ELLSWORTH, *Chief Counsel*
BENJAMIN S. COOPER, *Staff Director for the Minority*
HOWARD USEEM, *Professional Staff Member*
CRAIG GANNETT, *Senior Counsel, Minority*

FERC ELECTRIC UTILITY RESTRUCTURE

WEDNESDAY, MAY 10, 1995

U.S. SENATE,
COMMITTEE ON ENERGY AND NATURAL RESOURCES,
Washington, DC.

The committee met, pursuant to notice, at 2:21 p.m., in room SD–366, Dirksen Senate Office Building, Hon. Frank H. Murkowski, chairman, presiding.

OPENING STATEMENT OF HON. FRANK H. MURKOWSKI, U.S. SENATOR FROM ALASKA

The CHAIRMAN. Let me apologize for starting this hearing late. I made a commitment when I took this position that we were going to start on time. And so far, I have been able to live with it without exception.

So we are still on track for morning sessions. But obviously, we did not do as well for afternoon sessions.

You know, Betsy, having been involved in the process around here, that you look to staff for relief. And I am dictating to staff that we are not going to schedule my activities this tight anymore. I do apologize.

I have a short opening statement, which would be disappointing to some if I did not review it. So I will attempt to be brief.

First, I will call the hearing to order. This is the committee's first day of oversight hearings on the Federal Energy Regulatory Commission's proposal to restructure the electric utility industry and increase competition in the wholesale electric power market. We will have further hearings later this year after the industry has had a chance to digest the modest 500-page report.

I am also going to be particularly sensitive to the aspects of the issue of those power generating facilities which in effect, through one form or another, enjoy a subsidy. And we are going to be directing some attention to that area.

Now, as a general rule, I support measures which promote competition, as we all do. But as we all say, the devil is in the details. And certainly this is no exception. There are unlimited, unintended consequences associated with this proposal. And I am told it is called the Mega NOPR.

The Mega NOPR does not promote competition by eliminating Federal regulation. That is not necessarily satisfying. But instead, it does so by changing the way the FERC regulates the transmission of electric power.

While there are many appealing aspects to FERC's proposal, there are a number of unanswered questions, and that is why obviously we are starting this process.

I am interested in gaining a better understanding of the impact of the Mega NOPR on the structure of the electric utility industry, and particularly on the continued reliability of that industry. In short, we want to know if it will work. We also want to understand how it affects State public utility commissions' jurisdiction. I would also very much like to know how it can bring about fair competition when many of the players have special advantages, such as the Federal power marketing agencies and other utilities who are exempt from Federal taxes "for the time being."

Did you put that in?

[Laughter.]

The CHAIRMAN. I look forward to hearing a brief and concise explanation from FERC. Chair Moler will give us some readily understandable jargon concerning the complexities of the rulemaking, or at least give us some kind of a dictionary so we can pick up on some of it, particularly how it will affect consumers and the electric utility industry.

Unfortunately, Howard only wrote one page today.

So, Honorable Moler, please proceed and welcome to the committee. I trust you are comfortable. We look forward to hearing from you.

And there is room inside here, over here, for people who want to sit down. There is a whole table. We have the press at one side. Evidently, they are all conservatives or all liberals, because usually they split. So they are altogether today.

Come on in, folks, if you are inclined. Anybody outside that wants to come in? I do not know whether the fire marshal is out there or not.

[Pause.]

The CHAIRMAN. Please proceed.

STATEMENT OF ELIZABETH A. MOLER, CHAIR, FEDERAL ENERGY REGULATORY COMMISSION

Ms. MOLER. Good afternoon, Mr. Chairman.

It is indeed a pleasure to appear once again before this committee to discuss the Federal Energy Regulatory Commission's initiative on open access transmission and recovery of stranded costs.

As you have readily acknowledged, this is a major undertaking for us. We are dealing with a very complex subject matter. I will do what I can to explain what we are about in as simple terms as possible, but we all recognize the complexity of the industry that we are dealing with.

This rulemaking does represent the Commission's response to major changes that are underway in the electricity industry. We propose to respond to those changes in a comprehensive fashion and to deal with them comprehensively, and hopefully to provide regulatory certainty to all industry participants.

My testimony today will present an overview of our proposed rulemaking. As we discuss the proposed rule, I want to emphasize four important points. First, competition is coming to the electricity industry. Our goal is to ensure a fair and orderly transition from regulation to competition.

Second, all consumers should be able to reap the benefits of competitively priced generation. The Commission's proposed rulemaking works to ensure this result by requiring all jurisdictional utilities to file open access tariffs with the Commission. These tariffs will ensure that customers have access to wholesale suppliers on a non-discriminatory basis.

Third, utilities must be given an opportunity to recover the costs they incurred under the old rules to serve their customers. Open access and stranded costs are, in our view, inextricably linked. Open access will not succeed without a fair transition cost mechanism, and stranded cost recovery is appropriate only so long as open access is achieved.

Fourth, our existing regulatory structure is simply not adequate to meet the goal of a fair and orderly transition from regulation to competition. The question is not whether we need to amend our regulations. Rather, we must focus on how to change our regulations to achieve that goal.

There can be no question that the electric power industry is today an industry in transition to competition. As you observed, however, we are not comfortable entirely with promoting competition by eliminating regulation where there are still monopolies and where there is still a substantial potential or where indeed we find discriminatory practices in the industry.

That is what we are trying to deal with. We are dealing with monopoly facilities, meaning transmission facilities, and trying to encourage competition, particularly in the generation market.

Today, new generating facilities are being built at costs well below the average embedded costs of many vertically integrated utilities. Consumers are simply demanding access to these cheaper sources of power and suppliers are demanding access to customer markets.

Open access for transmission facilities is the key to bringing lower cost power generation to customers and to realizing the benefits of competition.

It is important for you and the other members of the committee to understand why we decided to proceed generically rather than proceeding on a case-by-case basis, as we have been doing.

The Energy Policy Act of 1992 encouraged competition in the wholesale, also known as the bulk power, market. In the Energy Policy Act, Congress gave the FERC authority to order transmission services on a case-by-case basis under section 211 of the Federal Power Act. We have taken our competitive mandate to heart and have aggressively implemented section 211.

However, competition in the generation sector and the demand for transmission access to get these benefits to customers have grown faster than anyone could possibly have expected. With the entry of significant, new third party power suppliers, we have heard an increasing number of complaints that those who own transmission facilities are discriminating against these third party power suppliers. Not surprisingly, the utilities are, in many instances, reluctant or simply unwilling to give up their monopoly control over the transmission facilities. This should not be surprising to anyone.

Last year, in response to complaints about discrimination, the Commission reexamined discrimination in transmission practices and announced a golden rule of comparability. Last May, for the first time we required that any utility voluntarily filing a transmission tariff must provide third parties with access to the transmission wires that is comparable to the way the utility uses its own system.

To date, we have applied this comparability standard on a case-by-case basis to a growing array of applications. We have applied it to applications involving transmission tariffs, which obviously raised discrimination issues, section 211 filings, mergers, and requests for market-based rates for new generation. All of these types of cases involve potential monopoly abuse of transmission wires.

The Commission's case-by-case efforts have met with some success. The comparability principle has become widely accepted in the industry. Several utilities have willingly filed open access tariffs offering comparable service.

However, we began hearing from industry participants that we should act generically to define the terms of comparable open access. Many in the industry were more than willing to provide open access, but they wanted to avoid litigating the specifics of the comparability requirement. They wanted guidance about what would be required for an acceptable comparability tariff so that they could move forward with their business decisions in an increasingly competitive world. The Mega NOPR is our response to those requests.

The majority of utilities still have not come forward with open access tariffs. When we issued the proposed rulemaking on March 29, only 17 of approximately 137 jurisdictional utilities, who own or control transmission facilities, had any form of comparable open access tariff on file with the Commission. Obviously, we have a long way to go. It has become clear that unless the Commission acts generically, the national transmission grid will become a patchwork of open access and non-open access systems, as well as a patchwork of systems offering different qualities and prices of services. There is undue discrimination inherent in this patchwork of transmission systems, because some customers can get the benefits of competition and others cannot.

We learned from our experience in the natural gas industry that it is best to deal with these issues up front in a comprehensive way so that this patchwork of open access and non-open access does not persist.

It has also become clear that the newly amended section 211, by itself, is not sufficient to remedy the undue discrimination and anti-competitive practices in the industry. Section 211 is a valuable tool to encourage competitive markets. However, there are delays inherent in the procedures for obtaining services under section 211, and for competitive reasons, some transactions must be negotiated quickly. Many competitive opportunities will be lost by the time a section 211 order can be obtained, no matter how we try. This means that customers served by open access systems can reap the advantages of such transactions. But other customers, whose utilities must go through a section 211 process, cannot obtain such benefits. The Commission therefore concluded that the patchwork of open and closed transmission systems could not be cured through the use of section 211 alone. Quite simply, generic action is required to remedy undue discrimination in the industry.

Section 211 will, of course, continue to be an important tool in developing nondiscriminatory access, particularly for non-public utilities that are not otherwise subject to our jurisdiction.

The Commission believes that its open access requirement, under the NOPR, and the use of Section 211 are complementary tools. However, there simply is no good substitute for readily available service under a tariff.

In my prepared testimony, I provide you with a summary of the key points of the proposed open access rule. In a nutshell, the cornerstone of the open access requirement is nondiscrimination.

All public utilities that own or control transmission facilities used in interstate commerce would be required to offer open access transportation over their facilities to purchasers and sellers of electric energy in the bulk power or wholesale market.

We do, of course, expect the third party users of the transmission system to pay a fair price for transmission and for ancillary services. We do not provide for retail wheeling. The Energy Policy Act prohibits us from doing so.

And finally, to ensure nondiscrimination, the proposed rule would require utilities to unbundle or separately price their wholesale services.

I want to emphasize that the open access proposal would not abrogate any existing power sales or transmission contracts. We propose to let the existing contracts run their course. Parties are free, of course, to renegotiate these contracts or to invoke notice provisions in the contracts to change power suppliers.

Let me now turn to stranded costs, the other critical element of the rulemaking. As I said at the outset, we believe it is critically important to address stranded costs at the same time we are requiring utilities to take part in an open access regime. You cannot have one without the other.

With the advent of competition, utilities are caught between the old rules of regulation under which they have made investments to serve customers and the new rules of competition under which they may lose those customers and see those investments stranded.

Competition and open access must go hand-in-hand with addressing stranded cost issues. Utilities, in turn, must realize that we will not be able to provide for recovery of stranded costs until we address the competition issues on a generic basis.

This is a package deal. We are trying to provide them a bridge between the old and new worlds.

Last July, we began to address this very difficult issue by issuing a notice of proposed rulemaking on stranded costs. We now have consolidated that stranded cost proceeding with the open access proceeding. We are now seeking comments on both open access and stranded cost issues.

Based on the comments we received to last July's proposals, we have come to some preliminary conclusions about stranded cost recovery that are spelled out in the proposed rulemaking.

First, and most importantly, we have found that utilities are entitled to a reasonable opportunity to recover legitimate, verifiable costs that could be stranded as a result of moving to a competitive industry.

We have made findings as to which types of stranded costs should be handled by the Federal Energy Regulatory Commission and which should be the responsibility of the appropriate State regulatory commission.

In conjunction with these findings, we set out standards for determining which facilities are used for transmission in interstate commerce and thus are subject to our jurisdiction, and which facilities are used for local distribution and thus are subject to State regulation.

We are, of course, at the proposal stage for our open access and stranded cost proposals. We have provided a 4 month comment period. Comments are due in August. We have also provided for reply comments 60 days later.

We look forward to a constructive, lively debate in the coming months. We appreciate your interest in our proposals. I would, of course, be happy to answer any questions you might have.

The CHAIRMAN. Thank you very much, Ms. Moler.

[The prepared statement of Ms. Moler follows:]

PREPARED STATEMENT OF ELIZABETH A. MOLER, CHAIR, FEDERAL ENERGY REGULATORY COMMISSION

Mr. Chairman and members of the committee, I am pleased to be here this afternoon to discuss the Federal Energy Regulatory Commission's recently proposed rulemaking on open access transmission by public utilities and recovery of stranded costs by transmitting utilities. This rulemaking represents the Commission's response to the major changes that are underway in the electricity industry. We propose to deal comprehensively with the changes and to provide regulatory certainty to all industry participants.

My testimony today will present an overview of our proposed rulemaking. As we discuss the proposed rule, I want to emphasize four important points.

First, competition is coming to the electricity industry. Our goal is to ensure a fair and orderly transition from regulation to competition.

Second, all consumers should be able to reap the benefits of competitively priced generation. The Commission's proposed rulemaking works to ensure this result by requiring all jurisdictional utilities to file open access tariffs. These tariffs will ensure that customers have access to wholesale suppliers on a non-discriminatory basis.

Third, utilities must be given an opportunity to recover the costs they incurred under the old rules to serve their customers. Open access and stranded cost recovery are inextricably linked: open access will not succeed without a fair transition cost mechanism and stranded cost recovery is appropriate so long as open access is achieved.

Fourth, our existing regulatory structure is not adequate to meet our goal of a fair and orderly transition from regulation to competition. The question is not whether we need to amend our regulations; rather we must focus on how to change our regulations to achieve that goal.

There can be no question that the electric power industry is today an industry in transition to competition. Today, new generating facilities are being built at costs well below the average embedded costs of many vertically integrated utilities. Consumers are demanding access to cheaper sources of power and suppliers are demanding access to customer markets. Open access for transmission facilities is the key to bringing lower cost power to consumers and to realizing the benefits of competition.

It is important for you to understand why we decided to address these issues generically with our proposed rulemaking, rather than proceeding case-by-case as we had been doing.

The Energy Policy Act of 1992 encouraged competition in the wholesale or "bulk power" market. In the Energy Policy Act, Congress gave the Commission authority to order transmission services on a case-by-case basis under section 211 of the Federal Power Act. We have taken our competitive mandate to heart and have aggressively implemented section 211.

However, competition in the generation sector—and the demand for transmission access to get these benefits to customers—has grown faster than anyone expected. With the entry of significant, new third party power supplies, we have heard an increasing number of complaints that those who own transmission facilities are discriminating against these new third party suppliers. Not surprisingly the utilities are, in many instances, reluctant or simply unwilling to give up their monopoly control over their transmission facilities. Discrimination in the type of transmission services being voluntarily provided has become a significant impediment to providing consumers access to lower cost power supplies.

Last year, in response to complaints about discrimination, the Commission re-examined undue discrimination in transmission practices and announced a "golden rule" of comparability. Last May, for the first time we required that any utility voluntarily filing a transmission tariff must provide third parties with access to the transmission wires that is comparable to the way it uses its own transmission system. We also required comparable services in response to requests for mandatory transmission services under section 211 of the Federal Power Act.

To date, we have applied this comparability standard on a case-by-case basis to applications involving transmission tariffs, mergers, requests for market-based rates for new generation, and other similar cases which involve potential monopoly abuse of transmission wires.

The Commission's case-by-case efforts have met with some success. The comparability principle has become widely accepted in the industry. Several utilities have willingly filed open access tariffs offering comparable service. We routinely set such proposals for hearing before an administrative law judge because the specifics of these filings are generally contested. By March of this year, we had 17 open access comparability hearings in progress. However, we began hearing from industry participants that we should act generically to define the terms of comparable open access. Many in the industry were more than willing to provide open access, but they wanted to avoid litigating the specifics of the comparability requirement. They wanted guidance about what would be required for an acceptable comparability tariff so that they could move forward with their business decisions in a competitive world.

The majority of utilities still have not come forward with open access transmission tariffs. When we issued the proposed rulemaking on March 29, only 17 of the approximately 137 utilities who own or control transmission facilities had any form of comparable open access tariff on file with the Commission. It has become clear that, unless the Commission acts generically, the national transmission grid will become a patchwork of open access and non-open access systems, as well as a patchwork of systems offering different qualities and prices of services. There is undue discrimination inherent in this patchwork of transmission systems because some customers can get the benefits of competition and others cannot.

It has also become clear that the newly amended section 211, by itself, is not sufficient to remedy the undue discrimination and anticompetitive practices in the industry. Section 211 is a valuable tool to encourage competitive markets. However, there are delays inherent in the procedures for obtaining services under section 211 and, for competitive reasons, some transactions must be negotiated quickly. Many competitive opportunities will be lost by the time a section 211 order can be obtained no matter how hard we try. This means that customers served by open access systems can reap the advantages of such transactions, but that customers who must go through a section 211 process cannot obtain such benefits. The Commission therefore concluded that the patchwork of open and closed transmission systems could not be cured through section 211. Quite simply, generic action is required to remedy undue discrimination in the industry.

All consumers should be able to reap the benefits of competitively priced generation. That is the primary reason the Commission has decided to require all public utilities to provide non-discriminatory transmission access to all legitimate wholesale buyers and sellers of electric energy, and to do so through open access tariffs. Section 211, of course, will continue to be important in developing non-discriminatory access, particularly by non-public utilities. The Commission believes that its open access requirement and the use of section 211 are complementary tools. However, there is no good substitute for readily available service under a tariff. That is what we seek to accomplish on a nationwide basis with our proposed open access rule.

Let me briefly describe the proposed open access rule.

The cornerstone of the open access requirement is nondiscrimination. All public utilities that own or control transmission facilities used in interstate commerce would be required to offer open access transportation over their facilities to purchasers and sellers of electric energy in the bulk power market. They would have to offer two types of tariffs, one that provides what is known as "point-to-point" transmission service and one that offers "network" service. They would also have to offer ancillary services. These are the types of services (such as scheduling and dispatching services) that have historically been provided exclusively by the host utility. They must also offer services necessary to maintain the reliability of the transmission system. We do, of course, expect the third party users of the transmission system to pay a fair price for transmission and for these other services. The proposed rulemaking proposes sample or pro forma tariffs that contain the minimum terms and conditions of open access, non-discriminatory service.

We do not provide for retail wheeling. The Energy Policy Act prohibits us from doing so.

To ensure non-discrimination, the proposed rulemaking would require utilities to "unbundle" their wholesale services. This means three things. First, when utilities engage in wholesale sales and purchases of electric energy, they would have to take transmission service under the same transmission tariff they offer to others. Second, they would have to price the transmission and ancillary services separately for each type of transmission service they provide. Third, they would have to rely on the same information network that their transmission customers rely on to arrange for the buying and selling of power. In other words, all transmission users would have access to the same information at the same time. The Commission has announced a technical conference to develop industry-wide standards for these information systems.

I want to emphasize that the open access proposal would not abrogate any existing power sales or transmission contracts. We propose to let the existing contracts run their course. Parties are free, of course, to renegotiate these contracts or to invoke notice provisions in the contracts to change power suppliers.

As I said at the outset, we believe it is critically important to address recovery of stranded costs at the same time we are requiring utilities to take part in an open access regime.

For those members of the Committee who may be unfamiliar with the stranded cost issue let me try to describe it and put it in context. Opening the transmission grid and providing wholesale access to cheaper power supplies will cause some utilities to incur what are known as stranded costs. For example, utilities with powerplants that produce high cost electricity are likely to lose customers on whose behalf those plants were built. The utilities may not yet have recovered their investment in these plants. Utilities also have obligations to buy power from many third party suppliers under the requirements of the Public Utility Regulatory Policies Act of 1978. The costs of power under these so-called "qualifying facilities" contracts in many cases are also in excess of current market prices for new power supplies.

With the advent of competition, utilities are caught between the old rules of regulation and the new rules of competition. Competition and open access must go hand-in-hand with addressing stranded cost issues. Utilities, in turn, must recognize that we will not be able to provide for recovery of stranded costs until we address the competition issues on a generic basis. We are trying to provide them a bridge between the old and the new worlds.

Last July, we began to address this very difficult issue by issuing a proposed rulemaking on the subject. We have consolidated that stranded cost proceeding with the open access proceeding. We are now seeking comments on both open access and stranded cost issues.

Based on the comments we already received to last July's proposal, we have come to some preliminary conclusions about stranded cost recovery.

First, and most importantly, we have found that utilities are entitled to a reasonable opportunity to recover legitimate, verifiable costs that could be stranded as a result of moving to a competitive industry. This includes costs that could be stranded if existing customers obtain wholesale transmission in order to reach another supplier, or if a state implements a retail wheeling or direct access regime. We have set out standards for which types of stranded costs should be handled by the Federal Energy Regulatory Commission and which should be the responsibility of the appropriate state regulatory commission.

For stranded costs that result from existing wholesale power customers obtaining wholesale wheeling as a result of the open access requirement or a section 211 order, we would be the forum for recovery. We would also be the primary forum for resolving any stranded costs that result from municipalization where former-retail customers become wholesale power customers. For stranded costs that result from retail wheeling, the proposed rule concludes that states should be the exclusive forum for recovery. The only exception would be where the state commission has no authority to provide for recovery at the time retail wheeling is required.

In conjunction with these findings, we set our standards for determining what facilities are used for transmission in interstate commerce, and thus are subject to our jurisdiction, and what facilities are used for local distribution, and thus are subject to state regulatory jurisdiction. This particular proposal is based upon a comprehensive analysis of the relevant case law that has developed under the Federal Power Act. The Commission enunciated several functional and technical factors that it will look at in making this determination.

Our jurisdictional proposal deals with issues that have not been previously addressed. But we believe it is essential to do so for two reasons. First, the standards will determine what facilities are subject to the Commission's open access requirement. Second, the definition of local distribution will permit states to retain jurisdiction over some of the facilities used in retail wheeling so that retail stranded costs can be recovered under state authority. While the Commission believes states have other mechanisms for assessing retail stranded costs, state jurisdiction over local distribution will assure that states have a means to address stranded costs that are caused by retail wheeling or a direct access regime.

At the same time we issued our open access and stranded cost proposal, we issued two other orders. The first, which I already mentioned, asked for comments and announced a technical conference on the type of real-time information networks—or "RINS"—we would require of transmitting utilities. The second was an order addressing how we plan to coordinate transmission access cases now pending before us with our final rule on open access. Copies of all three—the proposed open access and stranded cost rulemaking, the RINS Notice, and the case-specific order—have been furnished the Committee with this testimony.

We are, of course, at the proposal stage. We have provided a four-month comment period on the open access proposal. We have also provided for reply comments. We look forward to a constructive, lively debate in the coming months.

We appreciate your interest in our proposals. I would, of course, be happy to answer any questions you might have.

The CHAIRMAN. You spoke of the golden rule.

You also stated on page seven, "Quite simply, generic action is required to remedy undue discrimination in the industry."

It appears that the golden rule's application might be, "He who has got the gold, rules."

Ms. MOLER. Not surprisingly, I do not view it that way.

The CHAIRMAN. I see. That is just the way——

Senator BRADLEY. Maybe I came in late.

The CHAIRMAN. You did.

[Laughter.]

The CHAIRMAN. How do you see it, then? It sounds a little dictatorial if I just read that one sentence out of context, and I am sure you would agree with that.

Ms. MOLER. If you read one sentence in the rulemaking, it would be easy to read it out of context. I do agree with that, certainly.

We think it is important to have an open access regime across as broad a spectrum of utilities as we have jurisdiction over. If we proceed case-by-case, the process is going to be interminable. We would have to initiate individual section 206 proceedings against all of the utilities.

There is no doubt in my mind that unless we take generic action, that people who are frustrated in getting new transactions to occur and thwarted by utilities who simply will not let them get through their fence, that they will come to us.

They will either file section 211 applications, or else they will come to us, as they have in a couple of instances, and file petitions and say, "You, FERC, are not doing your job to remedy undue discrimination."

So we see no way to get from here to there in a timely fashion without taking generic action.

The CHAIRMAN. I am told that not very many petitions have been filed as of this date. Is that correct?

Ms. MOLER. We have—under section 211, we have had 19 petitions filed. I believe that is the current number. We also have had a number of individual complaints where, in the case that I mentioned last May, for example, the people who felt thwarted complained in an individual transmission case about undue discrimination.

So you cannot simply say how many section 211 filings have you had and how many complaints have you had. You really have to look at the landscape and figure out what you are seeing there.

The CHAIRMAN. The information I have suggests that the NOPR reports that of the 328 utilities subject to your jurisdiction, 137 own or control transmission facilities. And of the 137, 17 already have a pending or in-place open access transmission tariff.

Ms. MOLER. That is correct. Those tariffs vary a lot. Some of them are really good open access tariffs, and some of them are really bare bones.

The CHAIRMAN. And you propose, simply by ruling, incorporating the rest.

Ms. MOLER. Yes, sir.

The CHAIRMAN. Because they do not—why do the 17 voluntarily comply and the rest have not?

Ms. MOLER. That is a good question. Some of them have needed permission from us to do things. And in the course of seeking that permission, we have had complaints about discriminatory practices on their behalf.

Some utilities, for example, have subsidiaries where they want to sell generation from new powerplants at market-based rates.

The only way that we know how to make sure that they cannot use their transmission wires or otherwise discriminate in favor of themselves is to require them to adopt an open access tariff.

We have had merger cases, for example, where we have had complaints about discriminatory practices. And we have taken the open access principle and applied it in merger cases.

There is a growing list of circumstances in which the third party power suppliers are coming to us and complaining that a utility is abusing its monopoly power over its transmission wires. And we are having to deal with them.

The CHAIRMAN. Now, does the NOPR go too far in trying to impose one model or a single model for the industry restructuring, one perhaps which could be inhibiting the regional efforts to promote competitiveness based on different models? I am talking about California, Wisconsin, Massachusetts, Texas.

Ms. MOLER. We are at the proposal stage. I mean, this is a proposal—and we are seeking comment on it. We want to encourage regional models and regional planning. California, Wisconsin, New York——

The CHAIRMAN. Massachusetts.

Ms. MOLER [continuing]. Massachusetts, the States you mentioned and a growing list of other States, are asking themselves whether they should go even further than we are contemplating doing and whether they should have retail access or retail wheeling, whether large industrial consumers should be able to buy their power directly and in many cases perhaps skip the distribution system.

We do not deal with that issue in the NOPR at all. You have prohibited us from dealing with that issue, I would say at this point, thankfully. So we are not dealing with retail wheeling issues at all.

I also believe that there is a benefit from our doing an open access regime at the wholesale level, because you will be able to get a lot of the benefits of open access, and a lot of the competitive benefits can come about while stopping short of retail wheeling.

The CHAIRMAN. So as far as you are concerned, the jury is still out on the single model. You are willing to evaluate the considerations of the particular companies that would be affected a little differently than others. That is what you infer.

Ms. MOLER. We have—yes, we are. We have also said we will provide a starting point for an open access tariff.

The CHAIRMAN. Okay.

Ms. MOLER. That is what our pro forma tariffs are. We have also told the utilities: If you then want to come in and propose something other than our one-size-fits-all tariff, fine. Great. Come in. You know, come up with something that suits you individually. We will be happy to look at it.

We know it is not a one-size-fits-all industry, but we had to have a starting point.

The CHAIRMAN. Well, I am sure that many in the audience are pleased to hear that. I do not know how long it would take to put in place.

Another question. I have just one more, and I am going to turn it over to Senator Bradley.

The Mega NOPR is modeled, as I understand it, after FERC's open access transmission covering the natural gas industry.

Ms. MOLER. To a certain extent, yes.

The CHAIRMAN. And, you know, we have always had problems when we were differentiating between different types of energy and the cost transmission and so forth because of the physics of electric transmission vis-à-vis the physics of natural gas transmission.

As I understand it, the Mega NOPR open access requirement automatically goes into full effect for every electric utility 60 days after the rule is final.

Ms. MOLER. That is our proposal.

The CHAIRMAN. And until you do it, you do not really know if it is going to work or if it is going to have all the bugs out of it. And once you do it, there is no turning back after day 60.

What happens if the system does not work as the FERC and economists perhaps envisioned? And what happens to the reliability? We are both in trouble.

Ms. MOLER. We do not believe there will be a reliability issue in implementing this NOPR. We have dealt comprehensively with the types of services that utilities will now provide and will be providing in the future to make sure that the system still works.

We recognize and we talk about it extensively in the NOPR. We are working actively with the North American Electric Reliability

Council on this issue. They were in my office as recently as last week.

Nobody has said to us at the present time, "Hey, you know, open access is unreliable."

So we just do not see that as a problem.

We also think that by providing a starting point and then giving the utilities an opportunity to tailor make an open access system, that it will be a very efficient implementation process.

We learned on the natural gas side, when we did 2 years of individual cases and over 10,000 pages of individual orders, that there has to be a better way to do this that will work faster and be cheaper and more efficient for the agency to implement it. That is why we are proceeding as we are.

The CHAIRMAN. Well, it will be interesting for the record to note that when we have our next hearing, we are going to ask the utilities to respond to the same question relative to what is the risk, what happens if it does not work, and what is the likelihood?

My last question covers the automatic imposition to impose access 60 days after issuance of the final rule, which would result in an irrevocable change, effectively rendering moot any legal challenge.

The question is specific, and it is: Is the intent, scramble the eggs so that even if the Mega NOPR is held to be unlawful, it is impossible to reverse the changes?

Ms. MOLER. No.

The CHAIRMAN. Okay.

Ms. MOLER. If I may elaborate?

The CHAIRMAN. Please do.

Ms. MOLER. Our purpose in developing the tariffs is to take our experience that we have to date at developing these tariffs, take the best of the tariffs we have seen, and put them out for comment.

We are seeking comments on those proposed tariffs, as I said in my prepared testimony, so that we can respond to industry desires to provide them a more certain regulatory environment.

All of our initiatives that go into effect are later reviewed by the courts. That is just a fact of life that we deal with, and that is well accepted in the industry every day.

But our purpose with the tariffs is to give them as good an answer as we possibly can in defining a tariff that, if they adopt it, will be fine with us, and life can go on.

The CHAIRMAN. Thank you. There are going to be further questions that we will submit to you, but these are the priorities that I had. And I want to thank you very much.

I am going to call on Senator Bradley, who undoubtedly has a statement and questions as well. And I want to thank you.

If you will excuse me, Senator Bradley will Chair the balance of the hearing. We appreciate your professional presentation and the manner in which you have handled some tough questions. And we will give you some more tough questions in a written form.

Obviously, we will both share in the input from those that may have a little different explanation or anticipated concerns about the FERC proposal. Thank you very much.

Ms. MOLER. Thank you, sir.

The CHAIRMAN. Thank you, Senator.

Senator BRADLEY. Thank you very much, Mr. Chairman. You have given me an awesome responsibility here to be the Chair for the remaining of this hearing.

The CHAIRMAN. I am sure you are big enough for it.

Senator BRADLEY. And I thank you very much.

[Laughter.]

Senator BRADLEY [presiding]. Ms. Moler, let me say first that I am very pleased and supportive with the direction of the NOPR. I think that the idea of spreading the benefits of competition as broadly as possible is enormously important. And that is how I construe the actions that you are about to take.

I simply want to applaud that and say that as one Senator, I am prepared to work with you to try to figure out how we can achieve it even more.

Ms. MOLER. Thank you, sir.

Senator BRADLEY. In your statement, you indicate that you expect third party users of transmission are going to pay a fair price for the transmission services.

And, of course, the $1,000 question is: Well, what is fair price? I wondered if you had any thoughts or could describe how you would arrive at a fair price?

Ms. MOLER. Just as the industry is in transition, our historic way of doing rate making for transmission services is also in transition. We have had a tried and true postage stamp rate, one-size-fits-all proposal that has worked well for years.

The industry complained to us about that and said, "Please, you know, the world is changing. Give us some more flexibility."

We deliberated about that for a year and then last year announced a transmission pricing policy statement that gave the industry considerable pricing flexibility.

They, I think, are still digesting that and trying to figure out really what best suits their systems and how they solve some problems that are very difficult to solve because electrons do not flow where we lawyers want them to flow under contracts.

They have all the pricing flexibility in the world, but we do not have any proposals under the policy statement, and we stand ready to make good on the promise of giving them pricing flexibility. In the meantime, though, we have issued the NOPR.

So again, in our effort to figure out something that works and put something in place and do it in an efficient, cost-effective manner, which we are all trying to do these days as bureaucrats, we have said, "These are our proposed tariffs. Shoot holes in them if you will. This is the way we will figure your initial rates under the traditional postage stamp cost of service kind of rate making proposal. We will put those into effect immediately. And if you do not like the way we are proceeding, come on in and file your own proposal."

But in the meantime, we will have an open access network up and running, if you will.

Senator BRADLEY. So you are essentially assuming that your tariff schedule will provide the fair rates.

Ms. MOLER. Yes. And we are seeking comment on that.

Senator BRADLEY. And then you are open to comment as to whether they are unfair.

91-929 O - 95 - 2

Ms. MOLER. Absolutely.

Senator BRADLEY. I see. And then you would be responsive and try to assess the validity of those comments.

Ms. MOLER. That is what a rulemaking process is all about.

Senator BRADLEY. Okay.

Ms. MOLER. There are some other kinds of services that are not traditionally transmission services, that are really sort of the nuts and bolts of the way you make the transmission network work. They are called ancillary services. They are the most complex stuff we deal with.

It is difficult to quantify how much it costs the utilities to provide those services. We have put out a proposal on how to price ancillary services as well.

And we are sincere about seeking input, particularly on that issue, because it is something we do not have much experience with pricing separately.

Senator BRADLEY. So you are stepping off into the new world, but you want people with you.

Ms. MOLER. That is always best. And we welcome you.

Senator BRADLEY. Under the fair price scheme, do you expect the owners of transmissions lines to have the same or more net income from these facilities than they would otherwise have?

Ms. MOLER. Senator, I do not know that I know the answer to that question. There is an issue as we go about deciding what is jurisdictional to us and what is jurisdictional to the States, whether utilities will seek to move more of their facilities into FERC's jurisdiction and out of State jurisdiction. That is something that is a difficult issue to deal with.

Historically, we have had about 10 percent of utilities' revenue requirements under FERC jurisdiction, and about 90 percent of them are under State regulatory commissions. We are just peanuts in that sense, except for some limited number of utilities that are wholly wholesale.

So historically, we have been ten percent. We really do not think that the NOPR will be responsible for changing that. But we will find out.

Senator BRADLEY. But if the current owners receive more income than they would have originally expected from the fair price, why is that appropriate? How could it be appropriate that they would end up with more income?

Ms. MOLER. The circumstance under which they might conceivably end up with more income would be under some sort of incentive rate plan, where if they do a better job and can meet some benchmarks and so forth, that we will let them benefit from that.

We are exploring incentive rate plans in this policy statement I was talking about.

Senator BRADLEY. So that one should not assume more income a priori is a bad thing.

Ms. MOLER. No.

Senator BRADLEY. If there could be more income, would not this new income mitigate against the concerns about the utilities' investment in plants that might be stranded or rendered unprofitable by a more competitive power industry?

I mean, on the one hand, you have a stranded investment; on the other hand, you are going to get more income. Is there any kind of offset there?

Ms. MOLER. Transmission assets are depreciating facilities; and it costs a whole lot more to build a new transmission line than an old line.

Some people have advocated that you simply write up these transmission assets to a market-based rate and take the extra cash and write down your generating assets.

We only have jurisdiction over a part of the transmission assets, and we cannot make sure that the money that you take out of this set of pockets will necessarily go to the other pocket.

Senator BRADLEY. So you can make sure that they get more money, but you cannot make sure that they put it over on the generating side.

Ms. MOLER. We have not been terribly enthusiastic about those proposals. But if somebody comes to us with a plan where they can guarantee that is going to happen, we are not going to shut the door on them.

Senator BRADLEY. Would you say that, as you move to a competitive transmission system, that ultimately you will have to move to a competitive generating system?

Ms. MOLER. I personally think it is the other way around. And indeed, one of the proposals in the NOPR is that we codify that new generation should have market-based rates.

And I can see generation being done at a market price, rather than transmission, which to me is just a good old-fashioned monopoly.

But that could change as well. We are now exploring, on the gas side, where we have had a very robust open access regime, whether that theory is still true, whether they still are really natural monopolies or whether we can do something for a certain aspect of gas pipeline facilities, where they do not have any monopoly power.

Senator BRADLEY. Well, so tell me where are you on that? This is not a natural gas hearing, but where are you on your thinking with regard to that point on natural gas?

Ms. MOLER. We wrote our thinking down on a piece of paper. We put it out for comment—that is always dangerous, but it is what we do.

We put it out for comment and said, you know, "Where should we allow market-based rates and where should we not?"

Comments came in a couple weeks ago. We are reading those comments.

Some people are making the case that there are certain aspects of this sort of traditional business, like new storage for example, that while it has traditionally been considered transmission, we should no longer insist that it be at traditional cost-to-service rates, that we can allow market-based rates there. For example, some believe prices for "hub services" should be market based.

Hubs are where a lot of pipelines get together and gas comes in flows and flows out in many directions. The question is whether they are natural monopolies or whether we can allow market-based rates for hub services, things of that sort. And we are looking at those issues. These are second generation open access issues.

Senator BRADLEY. But it is amazing how far the whole thing has moved in just a few years, would you not say?

Ms. MOLER. Yes.

Senator BRADLEY. Moved in a predictable direction, too.

I understand that FERC has open access transmission tariff proposals on file. There are about 17 of them, as I understand, out of 137 regulated by FERC. Have you approved any of them?

Ms. MOLER. We have some that are in effect subject to refund while they go through the hearing process. And we have a couple of old open access tariffs that we approved before this rulemaking. So there are, I do not know, four or five others that are not in our current process.

Senator BRADLEY. Well, could you provide for the record summaries of the proposals that have been approved by FERC?

Ms. MOLER. Yes, sir.

Senator BRADLEY. Okay. Now we get to the chairman's favorite subject, the golden rule of comparability. It is kind of in the eye of the beholder, is it not? The basic question is: Could comparable rates be discriminatory?

I mean, obviously, you want a utility to provide access and rates for third party users so the grid that are comparable to the way the utilities own uses of its own transmission lines are. But could you imagine a circumstance where this becomes discriminatory?

For example, I mean, could not a utility, which is vertically integrated, shift its cost and income sources around in ways that might make comparable treatment to third party users very expensive to the third parties? In other words, you could manipulate your books.

You shift it up, down, around, out of this, into that, so that you are paying the same thing on this area, but because you are vertically integrated, you have managed to disadvantage those who would be competing with you.

Ms. MOLER. There are those who believe that the only way we can possibly get an effective, truly nondiscriminatory competitive world is to require the utilities to turn themselves into several companies and divest themselves, spin down, spin off, whatever you need to do from a tax point of view—you know more about that than I do.

We have not proposed to require to do that in this rulemaking. This rulemaking is already a very significant controversial undertaking.

Senator BRADLEY. Well, I agree with that.

Ms. MOLER. And we did not go that far. But I am sure we will have a lively debate, and we ask questions, you know, whether we should go further.

Senator BRADLEY. Well, in the NOPR, you state, "Because the traditionally vertically integrated utilities still favor their own generation, if and when they provide access to third parties, barriers continue to exist to cheaper, more efficient generation sources."

That is page 45.

Have you done any analysis of these other possibilities for market discrimination?

Ms. MOLER. I am frankly not sure I understand your question. We have looked at some industry structure issues. We are trying to figure out whether it is necessary to separate generation transmission and distribution——

Senator BRADLEY. Right.

Ms. MOLER [continuing]. In order to avoid favoritism and cross subsidies. We are actively exploring that issue. We do not think we have to, but I guess our analysis will come in response to comments on the proposed rule.

Senator BRADLEY. My question—I guess another way to put it is: How does the statement that I just read from page 45 of the NOPR—and again, let me read it to you.

It says, "Because the traditionally vertically integrated utilities still favor their own generation, if and when they provide access to third parties, barriers continue to exist to cheaper, more efficient generation sources."

In other words, if you are a small source, you are disadvantaged against the large vertically integrated.

And how does that compare to what you say on page nine, which is, "The Commission believes that the open access requirement will eliminate the transmission market power of public utilities"?

On the one hand you seem to be admitting that no, it is still going to be there. It is still going to be a problem. On the other hand, you are saying that you really think that it is much less of a problem and that you can solve it.

Ms. MOLER. The quotation from page 45 is a part of our description of why we believe we need to act.

Senator BRADLEY. I see.

Ms. MOLER. And it sets forth the problem. It is one of many pages that sets forth the problem we are dealing with. And our proposal to remedy that problem is open access.

Page nine——

Senator BRADLEY. I see.

Ms. MOLER [continuing]. You have the answer to the question, an analysis of what is going on.

Senator BRADLEY. Before the question. Right.

Ms. MOLER. On page nine, you have this summary of the answer to the problem.

Senator BRADLEY. I see. So basically the—I see. So basically I read the summary and thought there was a problem. But you had answered it in the summary when it arose in the question 45 pages later. This just sounds like FERC.

[Laughter.]

Ms. MOLER. Actually, people tell me it is written more eloquently than most of our stuff.

Senator BRADLEY. I am sure.

Ms. MOLER. Most of it is even worse.

Senator BRADLEY. Having benefitted from your prose for many years, if you wrote it, I am sure it is the best.

Let me ask: Do you have a time estimate, how long it will take for FERC to implement the rule based upon this NOPR?

Ms. MOLER. In a word, no.

Senator BRADLEY. So you do not have any idea about how long it will take the industry to implement.

Ms. MOLER. Well, once we tell them what they have to do, we are pretty sure we can make that happen. The question really is—we get comments in August, reply comments in October.

I am sure it will be a huge, multi-thousand page record we will have to deal with. We will have to figure out, you know, where we have done well, where we need to change our proposal, what problems have cropped up.

We fully anticipate that we will have to have some technical conferences on some issues. We are going to have to have a dialogue with our brothers, the State regulators, so that we make sure that we understand where we are and where our jurisdiction should end and where they think theirs should begin.

There are a bunch of issues we will have to deal with, and we will just have to sort through them. That is what we do.

Senator BRADLEY. From October——

Ms. MOLER. On.

Senator BRADLEY [continuing]. On until—what is your—you know, you are planning your life.

[Laughter.]

Ms. MOLER. I have 4 more years on my term.

[Laughter.]

Ms. MOLER. I anticipate happily——

Senator BRADLEY. I do not. I do not have——

[Laughter.]

Ms. MOLER. And thank you for that, for my term, not yours. I cannot—unless I move, I cannot vote for you.

Senator BRADLEY. Right.

Ms. MOLER. I fully anticipate we will be well beyond before I leave.

Senator BRADLEY. Okay.

Ms. MOLER. And I just do not know—I mean, it may be that people are going to tell us that this is a beautiful thing, and you should do it tomorrow. And if so, we will.

They may pose some problems that we have to solve. That was our experience on the gas side. We solved many problems after we got the comments and before we issued the final rule.

Senator BRADLEY. They basically responded to it.

Ms. MOLER. That is why you go through this process.

Senator BRADLEY. Right. And I think that it is, in this kind of matter, if everybody has basic acceptance of the premise that we are moving in more competitive markets broadly shared, benefits, blah, blah, blah, it is easy to work through them; not easy, but it is possible to work through them.

Ms. MOLER. And I do not believe that everybody will accept the premise.

Senator BRADLEY. I know.

Ms. MOLER. I think they are the ones who are going to fight us.

Senator BRADLEY. Right. But I encourage you to, and I think that those that fight really have the burden of proof, and ultimately it is a public burden. It is not a regulatory burden.

The NOPR says repeatedly that there is no reason to abrogate existing contracts. That is fairly straightforward. However, the NOPR would allow the public utility a claim on expiring contracts if the utility can prove that it had a reasonable expectation that the wholesale contract would have been renewed and service continued.

So what does that mean?

Ms. MOLER. What you are——

Senator BRADLEY. And how do you arbitrate between these two?

Ms. MOLER. What we are dealing with in that section is how do we quantify stranded costs. I sell power to you. I have forever. You have the ability now, at the expiration of our contract 3 years from now, to buy from Gene. You have not told me you are planning to do that. We were involved in a planning process. You have not given me any sort of advance notice.

If you leave and I have a powerplant that is worth $1 billion, did I have a reasonable expectation that you would continue to buy power from me, or should I have known upon expiration of the contract that you were going to leave and buy from Gene, who has a much cheaper source of power?

That is the issue. It is a much more pronounced and much more difficult problem, according to our record on the stranded cost NOPR, in a retail wheeling environment. In a wholesale environment, which is all we do, the initial record we got shows that the stranded costs, while they are significant, are not hundreds of billions of dollars——

Senator BRADLEY. Right.

Ms. MOLER [continuing]. Which they are in a retail environment. But what we are trying to do is get to expectations under the contract. And we will have to have hearings on those issues, because they will basically be prudence hearings on stranded costs.

Senator BRADLEY. Well, do you see—in this kind of arbitration process between these two, do you see any uncertainties that could be exploited?

And do you anticipate uncertainties that could be exploited to inhibit the development of more competition? You know, on the one hand, you are saying that the public utility is not going to abrogate existing contracts and then providing a way for them to.

Ms. MOLER. Well, we are saying that they can leave at the end of the contract.

Senator BRADLEY. And that is it. That is what you are saying.

Ms. MOLER. That is what we are saying.

Senator BRADLEY. So when you say "abrogate existing contracts," another way to say that would be "not renew existing contracts."

Ms. MOLER. That would be another way, yes. Lots and lots and lots of these contracts have notice provisions where you have to tell me you are going to leave, and then you are free to do so.

Senator BRADLEY. Right.

Well, again, let me tell you how much I respect what you are doing. This is not, as you say, going to be all dessert. But at the same time, it is a very important thing, I think, in terms of improving the overall efficiency and direction of the industry.

We have been through this now with natural gas for how long, about 8 years or longer?

Ms. MOLER. We actually finally finished implementing it only about less than 2 years ago.

Senator BRADLEY. It seems like 8 years.

Ms. MOLER. It seems like forever. It began——

Senator BRADLEY. From the first time we started these hearings.

Ms. MOLER. It began in about 1985, when the Commission began an initiative.

Senator BRADLEY. Right. That is 9 years.

Ms. MOLER. Ten years.

Senator BRADLEY. Ten years, yes. Well, with that background, we will be able to move much faster on electricity.

Ms. MOLER. Certainly.

[Laughter.]

Senator BRADLEY. And I am glad you are where you are, and I appreciate it. Thank you very much for coming before the committee today.

And as usual, you certainly helped me, and I think you helped other members of the committee as well.

Ms. MOLER. Thank you. We appreciate your words of encouragement. We really do.

Senator BRADLEY. I underline those words of encouragement. Go ahead, straightforward. Thank you.

[Whereupon, at 3:20 p.m, the hearing was adjourned.]

APPENDIX

RESPONSES TO ADDITIONAL QUESTIONS

FEDERAL ENERGY REGULATORY COMMISSION,
Washington, DC, July 11, 1995.

Hon. FRANK H. MURKOWSKI,
Chairman, Committee on Energy and Natural Resources, U.S. Senate, Washington, DC.

DEAR MR. CHAIRMAN: I appreciate your interest in the Commission's electric open access and stranded cost Notice of Proposed Rulemaking, "Promoting Wholesale Competition Through Open Access Non-discriminatory Transmission Services by Public Utilities, Docket No. RM95-8-000, and Recovery of Stranded Costs by Public Utilities and Transmitting Utilities, Docket No. RM94-7-001." Enclosed are my responses to your questions for the record of the May 10, 1995, hearing before your Committee.

I hope this information is helpful. If I can be of further assistance to you in this or any other Commission matter, please do not hesitate to contact me.

With best wishes,

Sincerely,

ELIZABETH A. MOLER,
Chair.

[Enclosure.]

RESPONSES TO QUESTIONS FROM SENATOR MURKOWSKI

Question 1. Sections 205 and 206 of the Federal Power Act prohibit undue discrimination. Since 1935, to determine whether or not a utility is engaged in undue discrimination, the Commission has looked at whether the utility is treating similarly situated customers differently. However, the Mega NOPR proposes to give a new definition to the term undue discrimination. It appears that the Commission has decided that undue discrimination also occurs if different utilities transmit power on differing terms and conditions, even though individually none is engaged in undue discrimination between its own customers. In fact, this claimed undue discrimination is taking place pursuant to FERC-approved tariffs.

Is this characterization of FERC's new definition of undue discrimination correct? What is the legal analysis and record evidence that supports the contention that all existing FERC-approved transmission tariffs are unduly discriminatory, even though individually none are unduly discriminatory?

Answer. The characterization of the Commission's definition of undue discrimination set forth in the question is not correct. As explained in American Electric Power Service Corporation (AEP), 67 FERC ¶61,168 at 61,490, clarified, 67 FERC ¶61,317 (1994), due to changing conditions in the electric utility industry, such as the emergence of non-traditional suppliers and greater competition in bulk power markets, the focal point of claims of undue discrimination has changed. In its role of protecting the public interest in this changing environment, the Commission must now be concerned not just with whether the utility treats one third-party transmission customer differently from another third-party transmission customer, but whether it provides itself with better transmission service than it provides to third-party transmission customers. Thus, if a public utility owns or controls interstate transmission facilities, but does not offer access to third parties on the same or comparable basis that the utility provides or could provide access to itself, it is acting in an unduly discriminatory manner. NOPR at 67-82.

The NOPR proposes that public utilities that own or control interstate transmission facilities be required to provide minimum non-price terms and conditions of service (as set forth in the tariffs attached to the NOPR) to all wholesale buyers and sellers of electric energy, including themselves. However, the Commission also proposes that these utilities may file tariffs offering different terms and conditions depending upon their individual circumstances (subject only to the minimum terms and conditions proposed in the NOPR). NOPR at 291-92.

We have explained in detail in the NOPR, including setting forth our legal analysis and record support, why we believe it is necessary to expand the Commission's traditional undue discrimination analysis in which we only addressed whether an individual utility is treating similarly situated customers differently. NOPR at 16-27; 63-65; and 67-82.

Question 2. In an open access environment, will those who have special legal, regulatory, or tax advantages have an unfair competitive advantage? Does the Mega NOPR have any provisions to adjust for this?

Answer. There is a large, and increasing, number of participants in the electric business today and their circumstances vary widely. Some participants may well enjoy tax and legal advantages as a result of business decisions, tax laws, State regulatory requirements or other factors beyond the Commission's jurisdiction or requirements. This is true today in the existing regulatory environment and will be true in the open access environment as well. The NOPR does not, and could not, address matters of tax policy or State law. However, it can ensure that all participants have the same opportunity to compete to provide generation service by fair rules governing access to the transmission grid.

Question 3. On page 8 of your written testimony you say that the Mega NOPR does not propose the abrogation of existing contracts. However, the Mega NOPR proposes to allow utilities to unilaterally file to amend a contract to include a stranded investment provision. Moreover, on page 133 of the Mega NOPR the Commission states that:

> However, there may be situations in which it would be contrary to the public interest to allow existing wholesale power or transmission contracts to remain in effect. *Accordingly. we invite comment on whether it would be contrary to the public interest to allow all or some of the above types of existing contracts to remain in effect.* (emphasis supplied).

Please explain.

Answer. The open access proposal contained in the NOPR would not abrogate any existing power sales or transmission contracts. This means that the rights and obligations of both buyer and seller, as specified in the existing contract, would be honored. As you correctly note, however, the NOPR does invite comment on whether our proposal not to abrogate contracts (i.e., to allow existing wholesale power or transmission contracts to remain in effect) would be contrary to the public interest.

The stranded cost proposal similarly respects the rights and obligations of the parties to existing contracts. When an existing contract explicitly provides for recovery of stranded costs, or precludes it, the terms of that contract will be honored. However, the NOPR further recognizes that there may have been "extra-contractual" expectations on the part of some utilities, who may have prudently incurred costs on behalf of existing customers whom they reasonably expected to remain on their systems beyond the contract term. The NOPR therefore proposes to permit the recovery of legitimate, prudent, and verifiable stranded costs associated with customers taking service under a limited set of existing wholesale power sales contracts—those contracts executed on or before July 11, 1994, that do not already contain exit fees or other explicit stranded cost provisions.

The NOPR encourages the parties to such power sales contracts in the first instance to renegotiate the contracts to reach a mutually agreeable resolution with regard to stranded costs. NOPR at 191. If such negotiations prove unsuccessful, the NOPR would permit a public utility (or its customer under the contract by complaint), at any time prior to the expiration of the contract, to file a proposed stranded cost amendment to the contract. In this regard, the Commission reaffirms its preliminary determination in the initial stranded cost NOPR, IV FERC Stats. & Regs. ¶32,507, 59 Fed. Reg. 35,274 (July 11, 1994), that it is in the public interest to permit public utilities to seek unilateral amendments to add stranded cost provisions if the contracts do not already contain exit fees or other explicit stranded cost provisions. NOPR at 192-93. Alternatively, a utility may seek recovery of stranded costs through transmission rates to the customer leaving the utility's system. To be eligible for recovery of stranded costs pursuant to a contractual amendment, or through transmission rates, a utility must meet the evidentiary and procedural criteria set forth in the NOPR for stranded cost recovery.

Question 4. The Mega NOPR automatically imposes identical open access transmission tariffs on all FERC jurisdictional electric utilities. These do not take into account any local or regional differences, including utility participation in power pools. Although utilities can subsequently file to amend their FERC-imposed open access tariff, given the slow pace of Commission action such a filing might not be acted on for many months, if not years. Should any special accommodation be made in the initial FERC-imposed open access tariff to account for local or regional differences, or for power pool operators?

Answer. At this time, we believe the proposed tariffs are sufficient to account for local or regional differences. However, we anticipate receiving comments on this issue and will carefully consider them in determining what additional flexibility can or should be included in any tariffs contained in the final rule.

The proposed pro forma tariffs are intended to provide the minimum terms and conditions of non-discriminatory transmission service. The NOPR is not intended to preclude utilities from proposing terms and conditions that go beyond this minimum, and that account for local or regional differences, as long as they do not permit the exercise of market power or undue discrimination. Regardless of the form of any pro forma tariffs ultimately adopted, utilities may file proposed tariffs prior to the final rule, or after the final rule, that differ from the pro forma tariffs. It is true, particularly in novel or controversial cases, that parties raise issues that may require lengthy hearings. However, FERC is required by law to act expeditiously (within 60 days) to review tariffs before a hearing process begins. A utility's proposed tariff could go into effect, perhaps subject to refund, within 60 days of the date the tariff filing was complete. As a result, any modification to the tariffs necessary to accommodate regional or local considerations can be accomplished in a reasonable period of time.

With respect to utility participation in power pools, the Commission acknowledged in the NOPR that power pools present complex issues that require special consideration. As the NOPR indicated:

> we intend to hold technical conferences with power pools to discuss implementation issues. After holding these technical conferences, and taking into account the comments received in the Open Access NOPR proceeding as well as in our pending Notice of Inquiry on Alternative Power Pooling Institutions, we will issue a supplemental order directing compliance for power pools.

NOPR at 291.

Question 5. Please identify by name each electric utility that owns or controls transmission facilities subject to the open access transmission requirement. Please identify by name each utility that is not subject to the open access transmission requirement and indicate why. Are there any states or large regions within which most, if not all, utilities are exempt from the open access requirement?

Answer. The proposed open access rule is being undertaken pursuant to sections 205 and 206 of the Federal Power Act (FPA). Sections 205 and 206 apply only to "public utilities," as defined in the FPA. Thus, persons that own and/or control facilities used for the transmission of electric energy in interstate commerce, and that are not Federal, State, or local entities, would be required to meet the requirements of a final rule requiring open access transmission. This includes most investor-owned electric utilities other than those located in Alaska, Hawaii, and the Electric Reliability Council of Texas (ERCOT). The proposed open access rule does not apply to all electric utilities or transmitting utilities that are owned by Federal, State, or municipal agencies, and cooperatives with loans administered by the Department of Agriculture's Rural Utilities Service (RUS).

The proposed rule estimated that about 137 transmission-owning investor-owned utilities operating control areas or participating in control areas (such as centrally dispatched pools or holding company pools) would be subject to the open access requirement of a final rule. This preliminary estimate was made to help define the approximate universe of entities that would be subject to a final rule for purposes of evaluating the reporting impact of the proposed rule. A review since the issuance of the proposed rule suggests that the number may be closer to 166. The additional utilities include some affiliates with transmission assets, smaller investor-owned utilities that do not operate control areas but may own transmission assets (including a few transmission-only companies), and some cooperatives with transmission assets that are jurisdictional because they are no longer RUS borrowers.

Attachment A identifies by name the 166 "public utilities" now identified as either meeting the control area criterion or as owning or controlling interstate transmission facilities. However, this list could change, depending on the final rule and the factual situation of each public utility. It should be noted that some affiliated utilities operating under holding company structures may file tariffs covering their combined integrated system.

There are regions throughout the country served by electric utilities not covered by the proposed open access rule. For example, public utilities subject to Commission jurisdiction under sections 205 and 206 of the FPA operate or participate as pools in 76 of the 140 control areas in the contiguous United States. The remaining 64 control areas (listed on Attachment B) are operated by utilities not subject to regulation under sections 205 and 206, and thus are not covered by the open access part of the proposed rule. These 64 control areas are scattered throughout the country. Many of them cover small geographic areas, such as municipalities. Alaska, Hawaii and Nebraska are the only states that have no utilities covered by the transmission access part of the proposed rule. Tennessee has very few utilities subject to the proposed rule because most of the state is served by the Tennessee Valley Authority (TVA) or the municipal systems TVA supplies. The Electric Reliability Council of Texas, representing a significant part of Texas, operates in a manner making some investor-owned utilities in Texas non-public utilities and thus also not subject to the proposed open access requirements.

Question 6. The Mega NOPR proposes to codify the Commission's determination in the recent *Kansas City Power and Light Company* case that wholesale electricity from new electric powerplants, but not from existing powerplants, can be sold at market-based rates. If the wholesale market is made competitive by the Mega NOPR's open access requirement, what is the policy argument against deregulating sales of wholesale power from existing powerplants? If sales of wholesale power from existing powerplants are not deregulated won't that place those utilities in an unfair competitive situation?

Answer. The *Kansas City Power and Light Company (KCP&L)* decision, proposed to be codified in the NOPR, was not a determination that sellers can automatically get market-rate authority for sales of generation from new electric power plants. The decision dropped only one of the showings—lack of generation dominance—that a seller must make in order to obtain market-based rate authority for sales from new plants. It did not eliminate the other showings. In addition, neither KCP&L nor the NOPR's proposed codification of KCP&L precludes a seller from obtaining market-rate approval for sales from existing electric power plants. This is discussed below.

The Commission's policy has been to permit market-based rates for generation from new or existing power plants if the seller can show that it and its affiliates lack market power in generation and transmission (or have mitigated such market power), there are no other barriers to market entry, and there is no evidence of affiliate abuse or reciprocal dealing. In the *KCP&L* case, the Commission dropped the requirement that applicants show a lack of market power in generation (also called generation dominance) if they are proposing to sell power from new plants. It did not drop the other requirements for obtaining market-based rates. The NOPR proposes to codify only the elimination of the requirement to show lack of market power in generation for sales from new units. Neither KCP&L nor the NOPR would preclude a utility from obtaining market-rate authority for sales of generation from existing capacity if it can show a lack of market power in generation and can meet the other criteria. Numerous utilities have indeed been able to meet all the criteria, including lack of generation dominance, and have authority to sell power at market-based rates from existing facilities.

In the NOPR, the Commission did not propose to change its existing criteria for market-based rates, either for sales from new capacity or for sales from existing capacity. However, it expressed the belief that the increased competition expected to result from the final rule may allow it to consider innovative approaches to authorizing market-based rates for generation from existing capacity. NOPR at 135. The Commission also stated that before dealing with market-based rates for existing generation, it must address certain threshold questions. Id. It then set forth those questions on which it solicits comments from interested parties. Id. at 135-36. The Commission's questions reveal its concern that wholesale sellers of existing capacity may still possess market power in an open access environment. The Commission will review the comments submitted in response to these questions before determining whether to drop the generation dominance requirement for existing generation, or to make other changes to its market-rate criteria.

Question 7. What costs would the Mega NOPR allow to be recovered from a departing wholesale customer? Are these all of the costs the utility has incurred with respect to that customer? If not, what happens to the unrecovered costs? Are they absorbed by the utility or its shareholders, or are they passed on to the utility's remaining customers? How will that occur? Will the utility have to file a rate case?

Answer. The NOPR proposes that the determination of recoverable stranded costs be based on a "revenues lost" approach. Under this approach, a utility would calculate a departing wholesale customer's stranded cost liability by subtracting (a) the competitive market value of the power the customer would have purchased from the utility (and the basic revenues from the transmission service) had the customer continued to take generation service under its contract from (b) the revenues that the customer would have paid the utility. NOPR at 215-16. As the Commission explained in the NOPR, the "revenues lost" approach in effect eliminates the need to list categories of costs that may be recovered. NOPR at 218-19. However, the NOPR also recognizes that there may be special categories of costs that are properly allocated to departing customers and that are not captured in the "revenues lost" approach. Nuclear decommissioning costs, for example, may not be reflected (or may not be fully reflected) in current requirements rates. The NOPR accordingly seeks comments on whether there are special costs that warrant some special consideration in the determination of stranded cost liability under a "revenues lost" approach, and if so, how they should be treated. NOPR at 219.

Under normal Commission practice, if a utility does not recover costs from a departing customer, and the costs were prudently incurred, the utility may seek to reallocate those costs to remaining customers in its next rate case. In its initial stranded cost NOPR, IV FERC Stats. & Regs. ¶32,507, 59 Fed. Reg. 35,274 (July 11, 1994), the Commission stated that it was not addressing in the NOPR who would bear stranded costs caused by a departing customer if the Commission finds that the utility had no reasonable expectation of continuing to serve that customer. It further stated that it anticipated a public utility would seek in subsequent requirements rate cases to have the costs reallocated among the remaining customers on the system. IV FERC Stats. & Regs. at 32,872. The Commission also discussed the situation in which a departing customer leaves a utility's system by using another utility's transmission system. It stated that in this situation it anticipated that any stranded costs would be reallocated to remaining customers in the utility's next requirements rate case, but sought comment on this issue. Id. The Commission did not further address these issues in its March 29, 1995 supplemental stranded cost NOPR, but did seek comment on what means ought to be available for the collection of stranded costs in the event that a seller establishes that it had a reasonable expectation of continuing to serve a departing customer, and the departing customer does not take unbundled transmission service from the former seller. NOPR at 206.

Question 8. In a theoretical sense, stranded costs also occur as a result of lower wholesale prices brought about by competition. Does the Mega NOPR provide for the recovery of competition-caused stranded costs? If not, what happens to those stranded costs? Are they absorbed by the utility or its shareholders, or are they passed on to the utility's remaining customers? How will that occur? Will the utility have to file a rate case? How might the passthrough of those costs affect rates?

Answer. The proposed stranded cost rule is not intended to insulate a utility from the challenges of a competitive wholesale generation market. Rather, it is intended to ensure only that prudently incurred costs associated with providing electric service under the pre-existing regulatory regime will be recovered. This honors the terms of the regulatory compact, extant prior to the widespread introduction of competition into the generation sector, whereby utilities were expected to plan for customers' generation needs and, in return, could reasonably expect to recover the prudently incurred costs of meeting those needs. As such, stranded costs are the cost of transition to the new competitive environment, in which all suppliers will have the opportunity to compete.

Question 9. The Commission's May 8, 1995 *West Penn Power* decision speaks to the issue of buy-down and buy-out of existing wholesale power purchase contracts. In it, the Commission states:

> Rather, we believe the appropriate action is to buy out or buy down such contracts. To facilitate such action, we clarify that if utilities are prudent in buying out or buying down existing power purchase agreements, whether or not with QFs, this Commission will permit the recovery in wholesale rates of a pro-rata share of the buy-out or buy-down costs.

Does *West Penn Power Company* indicate a Commission policy that wholesale power purchase contracts, including QF contracts, are eligible for stranded cost recovery? Will this be incorporated into the Mega NOPR's stranded cost provisions? What is the "pro-rata" share that will be eligible for stranded cost recovery? Can state commissions deny the subsequent passthrough in retail rates of these costs? If states do not allow their pro-rata share to be passed on in retail rates, what occurs? Will the recovery in wholesale rates go on all wholesale sales, be assigned to specific customers, added on to transmission rates, or be assigned to departing wholesale customers? What is meant by "recovery in wholesale rates"? What if a utility has no or virtually no wholesale customers? Will any of these costs be assigned to unbundled retail transmission, over which the FERC asserts jurisdiction? What does the phrase "prudent in buying out or buying down" mean and how will it be determined? Is it the prudence of how the utility buys out or buys down the contract, or is it the prudence of the contract? Will the utility have to file a rate case before it can pass through these costs?

Answer. *West Penn Power Company* does not address whether wholesale power purchase contracts, including QF contracts, are eligible for stranded cost recovery. However, the stranded cost NOPR proposes that any legitimate, prudent and verifiable costs incurred by a public utility could be recovered by the utility if it meets the evidentiary showings necessary for recovery. This would include costs associated with purchase power contracts, including QF contracts. It also would include buy-out or buydown costs. However, because the NOPR proposes a "revenues lost" approach (see response to question 7), this eliminates the need to distinguish categories of costs. If the utility satisfies the necessary evidentiary showings, a departing wholesale customer would pay a pro rata share of all wholesale costs incurred. These costs could be paid either through an adder to the departing customer's transmission rate or through an exit fee to its power sales contract. As we indicated in West Penn Power Company, permitting the recovery of buy-out/buy-down costs associated with QF contracts is consistent with Commission precedent allowing the recovery of prudently-incurred costs associated with the buy out or buy down of higher-priced fuel contracts in order to substitute currently available lower-priced fuel. Slip op. at 26.

Under Commission precedent, prudently incurred costs are costs that a reasonable person would have incurred based on the facts and circumstances known at the relevant point in time. Thus, even though it may have been prudent to enter into a contract at one time, if circumstances later change, it may become prudent to attempt to buy-out or buy-down that contract. A utility would have to file a rate case to seek recovery of such costs.

"Recovery in wholesale rates" means that the seller can recover such costs through a charge to wholesale customers. Normally this would be through rates for wholesale power sales (i.e., bundled generation and transmission sold to a wholesale customer). As discussed above, it is also possible that such costs could become part of stranded costs and recovered through unbundled transmission rates. Regarding the portion of buy-out or buy-down costs allocated to retail load, a utility would have to seek recovery from the state regulatory authority to recover these costs in bundled retail power sales rates. If the states do not allow recovery of the retail share to be passed through retail rates, the utility could try to seek a reallocation of those costs to wholesale customers. I cannot comment on whether the Commission would permit this kind of reallocation.

Question 10. The Mega NOPR notes the importance of recovery of stranded investment as part of opening up the wholesale market. In *Cajun,* the Court of Appeals questioned FERC's allowance of stranded cost recovery in a utility's transmission rates. If the Mega NOPR's stranded cost recovery system is held unlawful, how would that affect the implementation of open access transmission? What would the Commission do if it is held unlawful after open access transmission is implemented?

Answer. As the Commission stated in the NOPR, "[i]t is important to couple the Commission's open access rule with a rule ensuring recovery of all legitimate transition costs." NOPR at 10. The manner in which the Commission has linked stranded cost recovery with the successful transition to an open transmission access, competitively-priced industry was designed to address the concerns of the court in *Cajun.* In the NOPR, the Commission is providing the evidentiary record for addressing all of the *Cajun* court's concerns on a generic basis, and the opportunity for all participants in the electric industry to present evidence and arguments.

For example, the court in *Cajun* was concerned about whether Entergy's tariff allowed "meaningful" access to alternative suppliers. A fundamental purpose of the NOPR is to ensure the meaningful access to alternative suppliers that was identified by the *Cajun* court. NOPR at 167-68. In addition, in response to the *Cajun* court's holding that the Commission had failed to explain adequately its approval of Entergy's stranded cost provision, the NOPR provides a detailed explanation of why the recovery of legitimate stranded costs is critical to the successful transition of the electric utility industry to an open transmission access, competitive industry that will drive down the prices of electricity to consumers. As we note in the NOPR,

> we have gathered (and are continuing to gather) information concerning the magnitude of potential stranded costs; we have provided an explanation of the transitional nature of stranded costs; and we have explained the critical need to deal with these costs in order to competitive wholesale markets. We have also explained existing disparities in electricity rates and the consumer benefits that can accrue if we achieve fully competitive markets.

NOPR at 169.

On this basis, we have every reason to believe that the NOPR's stranded cost recovery provisions are lawful. However, if those provisions were held unlawful (whether before or after the implementation of open access transmission), the Commission would consider at that time the impact of such a determination on how to implement an open access transmission regime.

Question 11. What provision of the Federal Power Act gives the FERC the authority to require electric utilities to expand their transmission systems? Does that include the authority to expand transmission capacity even if is not required to serve native load? The Commission has asserted jurisdiction over the rates, terms and conditions of unbundled retail transmission. Does the Commission likewise assert the authority to order electric utilities to expand their transmission system for unbundled retail sales?

Answer. The Commission has the authority under sections 205 and 206 of the FPA to remedy any undue discrimination by public utilities with respect to the transmission of electric energy in interstate commerce and sales for resale of electric energy in interstate commerce. Thus, to the extent necessary to remedy any undue discrimination with respect to transmission in interstate commerce, we can direct public utilities to expand their transmission systems to offer wholesale transmission services. This includes the authority to direct expansion of transmission capacity even if the expanded capacity is not required to serve native load. However, the Commission does not have siting or certificate authority with respect to the construction of new facilities. Such activities are subject to State jurisdiction.

We also do not have authority to require public utilities to expand their systems to offer transmission services directly to an ultimate consumer. We have asserted jurisdiction only over the rates, terms and conditions of unbundled transmission services by public utilities to retail customers. Such unbundled retail wheeling may occur voluntarily or as the result of state or local action. We have taken no position on whether state or local entities have the authority to order retail wheeling.

Question 12. If a state public utility commission orders or allows retail wheeling, will it have to be performed pursuant to the Federal transmission standards established by the Mega NOPR? Is there any policy or legal reason why state public utility commissions should not be able to establish their own program of retail wheeling without having to conform to FERC's wishes?

Answer. As discussed above, the Commission does not have authority to order wheeling directly to an ultimate consumer. The Commission has not taken a position on whether states have authority to order retail wheeling. If retail wheeling occurs, as noted above, the Commission is obligated under the Federal Power Act to regulate the rates, terms and conditions of the resulting transmission by a public utility in interstate commerce.

Question 13. What is the demarcation between federal and state jurisdiction over transmission of electricity? Is any unbundled transmission exempt from FERC jurisdiction? Is any bundled retail sale of electricity subject to FERC jurisdiction? What is the line between transmission and distribution activities? Who has jurisdiction over transmission facilities that transmit both bundled retail power and wholesale power? Who has jurisdiction over transmission facilities that transmit both bundled retail power and unbundled retail power? Would you recommend a "Hinshaw" amendment for the Federal Power Act?

Answer. The line between Commission-jurisdictional transmission in interstate commerce and state-jurisdictional local distribution in the context of unbundled wheeling by public utilities is addressed in section G of the NOPR (Transmission/Local Distribution) (NOPR at 249-86). In the case of uubundled wholesale wheeling (i.e., delivery in interstate commerce by a public utility of electric energy from a third-party supplier to a purchaser who will then re-sell the electric energy to an end user), the Commission concluded that the case law supports the drawing of a "functional" line. Under this "functional" test, a public utility's facilities used to deliver electric energy to a wholesale purchaser are subject to the Commission's exclusive jurisdiction under sections 205 and 206 of the FPA, and a public utility's facilities used to deliver electric energy from the wholesale purchaser to the ultimate consumer are "local distribution" facilities subject to the rate jurisdiction of the state. NOPR at 251; see also NOPR at 282-83.

In the case of unbundled retail wheeling (i.e., delivery in interstate commerce by a public utility of electric energy from a third-party supplier directly to an end user), the Commission concluded in the NOPR that "some of the public utility's facilities used to deliver electric energy to an end-user may be FERC-jurisdictional transmission facilities, while some of the facilities used may be state-jurisdictional local distribution facilities." NOPR at 251-52. In particular, where a former bundled retail customer may need unbundled wheeling services from its previous public utility generation supplier as well as unbundled wheeling from one or more intervening public utilities in order to reach a distant generation supplier, we believe that the Commission "would have jurisdiction over all of the facilities used for the unbundled wheeling provided by the intervening public utilities." NOPR at 283. With regard to the facilities used to transmit energy from the transmitting utility in closest proximity to the end user (the former supplier of the bundled product), in most, if not all circumstances, some portion will be local distribution facilities. NOPR at 283-84. The NOPR proposes a combination functional-technical test to determine whether particular facilities are transmission or local distribution that will take into account the following indicators (NOPR at 285):

- Local distribution facilities are normally in close proximity to retail customers.
- Local distribution facilities are primarily radial in character.
- Power flows into local distribution systems, it rarely, if ever, flows out.
- When power enters a local distribution system, it is not reconsigned or transported on to some other market.
- Power entering a local distribution system is consumed in a comparatively restricted geographical area.
- Meters are based at the transmission/local distribution interface to measure flows into the local distribution system.
- Local distribution systems will be of reduced voltage.

A bundled retail sale of electricity (i.e., a retail sale of electric energy coupled with the transmission and local distribution of such energy to the end user) would not be subject to Commission jurisdiction.

In the case of facilities that are used to deliver both bundled retail power and bundled wholesale power, as we indicate in the NOPR, "we believe that the Commission and the States have jurisdiction to set rates for the services that are within their respective jurisdictions." NOPR at 251 n.306.

In the case of facilities that are used to deliver both bundled retail power and unbundled retail power, the facilities used to deliver the bundled retail power would be state-jurisdictional; some of the facilities used to deliver the unbundled retail power may be FERC-jurisdictional transmission facilities, while some of the facilities used may be state-jurisdictional local distribution facilities.

With the NOPR's clarification of the tests the Commission will apply to determine if facilities used to deliver unbundled electric energy are FERC-jurisdictional or state-jurisdictional, I see no need to consider a "Hinshaw" amendment to the Federal Power Act.

Question 14. What information will RINS require to be disclosed? Who will be required to disclose that information? Will any generation information be required to be disclosed? Will RINs put those who are required to disclose at a competitive disadvantage with those who are not required to disclose? Does RINs contain a reciprocity provision?

Answer. We will be addressing these questions in a separate proceeding concerning RINs initiated concurrently with the Open Access NOPR. See Real-Time Information Networks, Notice of Technical Conference and Request for Comments, IV FERC Stats. & Regs. ¶35,530, 60 Fed. Reg. 17,726 (1995). In issuing the notice, we suggested that the information that should be available on a RIN includes "transmission availability data, scheduling information, information on economic dispatch, system reliability conditions, service interruptions, and other information that parties might suggest." Notice at 4-5. We will also address in the RINs proceeding whether any generation information will be required to be disclosed.

In the Notice, we explained:

> The Commission is considering requiring each public utility (or its agent) that owns and/or controls facilities used for the transmission of electric energy in interstate commerce to create a real-time information network (RIN) to ensure that potential purchasers of transmission services have access to information to enable them to obtain open access transmission services on a non-discriminatory basis from the public utility.

Notice at 1.

The goal of RINs is to ensure that truly non-discriminatory open transmission access is provided. Unless all wholesale buyers and sellers have access, on a real-time basis, to information on transmission availability, transmission access will not be comparable to that which transmission owners provide themselves. Such access to information will therefore remove competitive disadvantages. The Commission anticipates constructive dialogue on this issue in comments and at the technical conference(s) held in the proceeding.

The proposed tariffs contain a reciprocity provision that ensures that a public utility providing transmission access to others can obtain similar service from its transmission customers. NOPR at 119-20. Thus, to the extent that an entity that is not a public utility receives transmission service from a public utility, as a condition of taking service under the proposed tariffs, it would have to offer similar service to the public utility, including making information available through a RIN.

Question 15. How will RTGs and power pools be integrated into open access transmission?

Answer. In the NOPR, we discussed RTGs and power pools. With respect to RTEs, we explained, in part:

> To the extent public utilities view the comparability requirement in our two RTG orders as a disincentive to joining an RTE, that disincentive would be mooted. All such utilities will be required to file tariffs. Moreover, we will continue to provide substantial latitude for innovative pricing proposals by an RTG, as indicated in the Transmission Pricing Policy Statement.
>
> Some transmission users might conclude that the availability of comparability tariffs makes membership in an RTE less necessary. But, this conclusion would ignore the comparative benefit of a member having its needs planned for on a region-wide basis under an RTG instead of on a system-by-system basis.

NOPR at 137. In conclusion, we stated that "[t]he potential benefits of RTGs would in no way be undermined by the rules proposed in this Open Access NOPR." NOPR at 138.

As discussed above in the answer to question 4, the Commission has recognized that power pools present special implementation issues that the Commission will investigate through a technical conference in the near future.

Question 16. Given that actual transmission capacity varies depending on which generating plants are actually being used, as well as which loads are being served, will it be possible to make transmission capacity a freely tradeable commodity?

Answer. In the NOPR, we recognized that some services may be more reassignable than others. With respect to point-to-point transmission service, we explained:

> Point-to-point service involves a well specified right to transfer a given amount of power between specific points or across an interface under certain conditions. The transmission provider is operationally indifferent as to who wants to transfer the power that flows between those points. Thus, point-to-point service is well-suited to reassignment.

NOPR at 118.

With respect to network and ancillary services, we recognized that these services may not be reassignable and sought comments on this issue. NOPR at 119.

Question 17. What impact will the Mega NOPR's requirement that utilities have flexible point-to-point transmission have on transmission revenues? If it reduces revenues, would the lost revenue be eligible for recovery as a stranded cost? If not, would the utility's native load be responsible for the costs?

Answer. The NOPR's flexible point-to-point transmission requirement permits a transmission customer to change points of receipt and delivery on an as available, non-firm basis without incurring additional costs. The Commission has no intent, either by this proposal or any other proposal in the NOPR, to depart from its historical practice of permitting the recovery of all prudently incurred costs of transmission. Thus, there should not be any reduction in transmission revenues as a result of flexible point-to-point service. Nor would native load be responsible for any of the costs of providing such service to a third party.

If transmission revenues were to become part of stranded costs, it would not be because open access tariffs offer flexible point-to-point service. Rather, it would be in the unlikely event that a customer departs a utility's system by using someone else's transmission system to reach a new generation supplier, and the utility is unable to recover through an exit fee the resulting stranded costs (including transmission costs that would have been borne by the departing customer had it stayed on the utility's transmission system). The Commission then would have to decide whether to permit the utility to recover the transmission costs from the remaining customers. As discussed in the response to question 7, the Commission sought comment in the NOPR on the issue of stranded costs caused by departing customers who leave their supplier via another utility's transmission system.

Question 18. In *Otter Tail Power Company*, the Supreme Court held:

> "So far as wheeling is concerned there is authority granted the Commission under Part II of the Federal Power Act to order it, for bills originally introduced contained common carrier provisions which were deleted. The

Act as passed contained only the interconnection provision set forth in Section 202 (b). The common carrier provision in the original bill and the power to direct wheeling were left to the "voluntary coordination of electric facilities. . . ." *Otter Tail Power Co.* v. *United States,* 410 U.S. 375-76 (1973) (emphasis supplied) (footnotes omitted).

Does this decision bear on the Mega NOPR's open access transmission requirement? In the Energy Policy Act of 1992, did the Congress take a position on requiring utilities to file transmission tariffs of general applicability?

Answer. We fully explained our legal authority to remedy undue discrimination and to require non-discriminatory open access as a remedy for undue discrimination. NOPR at 16-28. In doing so, we discussed *Otter Tail* and explained that it "did not address whether the Commission can require transmission in fulfillment of its duty to remedy undue discrimination." NOPR at 19-20. We also discussed section 211 of the FPA, as amended by the Energy Policy Act of 1992. NOPR at 27-28. We explained that "[i]n amending section 211 Congress left unaltered the authorities and obligations of the Commission under sections 205 and 206 (similar to our authorities and obligations under sections 4 and 5 of the Natural Gas Act) to remedy undue discrimination." NOPR at 27 n.35.

Question 19. What is the Commission's position on the issue of functional and corporate unbundling of electric utilities? Does the Commission have the statutory authority to require corporate unbundling? Is corporate unbundling under consideration?

Answer. Functional and corporate unbundling are addressed in the NOPR at pages 94-102. We explained that:

> the proposed rule requires that a public utility's uses of its own transmission system for the purpose of engaging in [new] *wholesale* sales and purchases of electric energy must be separated from other activities, and that transmission services (including ancillary services) must be taken under the filed transmission tariff of general applicability. The proposed rule does not require corporate unbundling (selling off assets to a non-affiliate, or establishing a separate corporate affiliate to manage a utility's transmission assets) in any form, although some utilities may ultimately choose such a course of action. The proposed rule accommodates corporate unbundling, but does not require it.

NOPR at 94 (emphasis added).

[Attachment A]

Public Utilities that operate Control Areas or are within Control Areas operated by a Public Utility or their agent.

NERC region or subregion	NERC defined control area	Company name—NOPR jurisdictional
ECAR	AEP	Appalachian Power Company
		Columbus Southern Power Company
		Indiana Michigan Power Company
		Kanawha Valley Power Company
		Kingsport Power Company
		Kentucky Power Company
		Ohio Power Company
		Wheeling Power Company
	APS	Monongahela Power Company
		Potomac Edison Company
		West Penn Power Company
	CGE	Cincinnati Gas and Electric Company
		Union Light, Heat & Power Company
	CEI	Cleveland Electric Illuminating Company
	DPL	Dayton Power & Light Company a
	DLCO	Duquesne Light Company
	IPL	Indianapolis Power & Light Company
	KU	Kentucky Utilities Company
	LGE	Louisville Gas & Electric Company
	MECS	Consumers Power Company
		Detroit Edison Company
		Edison Sault Electric Co.
	NIPS	Northern Indiana Public Service Company

NERC region or subregion	NERC defined control area	Company name—NOPR jurisdictional
	OES	Ohio Edison Company
		Pennsylvania Power Company
	OVEC	Ohio Valley Electric Corporation
		Indiana-Kentucky Electric Corporation
	PSI	PSI Energy, Inc.
	SIGE	Southern Indiana Gas and Electric Company
	TE	Toledo Edison Company
ERCOT	CPL-WTU	Central Power & Light Company
		West Texas Utilities Company
MAAC	PJM	Atlantic City Electric Company
		Baltimore Gas & Electric Company
		Delmarva Power & Light Company
		Jersey Central Power & Light Company
		Metropolitan Edison Company
		PECO Energy Company
		Pennsylvania Electric Company
		Pennsylvania Power & Light Company
		Potomac Electric Power Company
		Public Service Electric & Gas Company
		UGI Utilities, Inc.
MAIN	CE	Commonwealth Edison Company
		Commonwealth Edison Company of Indiana
	CIL	Central Illinois Light Company
	CIPS	Central Illinois Public Service Company
		Mt. Carmel Public Utility Co.
	EEI	Electric Energy Inc.
	IP	Illinois Power Company
	MGE	Madison Gas & Electric Company
	UE	Union Electric Power Company
	UPP	Upper Peninsula Power Company
	WEP	Wisconsin Electric Power Company
	WPL	Wisconsin Power & Light Company
		South Beloit Water, Gas & Electric Co.
	WPS	Wisconsin Public Service Company
		Consolidated Water Power Co.
MAPP	IESC	IES Utilities Inc.
	IIGE	Iowa-Illinois Gas & Electric Company
	IPW	Interstate Power Company
	MP	Minnesota Power Company
		Superior Water, Light and Power Co.
	MPSI	Midwest Power Systems Inc.
	NSP	Northern States Power Company—MN
		Northern States Power Company—WI
	OTP	Otter Tail Power Company
NPCC NEPOOL.	NEPEX	Bangor Hydro-Electric Company
		Boston Edison Company
		Central Main Power Company
		Central Vermont Public Service Corporation
		Citizens Utilities Co.
		Connecticut Valley Electric Co. Inc.
		Green Mountain Power Corporation
		New England Hydro Transmission Corp.
		New England Hydro Transmission Electric Company
		United Illuminating Company
		Vermont Electric Power Company
		Vermont Electric Transmission Company
		Connecticut Light & Power Company
		Western Massachusetts Electric Company
		Public Service Company of New Hampshire
		Holyoke Water Power Company
		Holyoke Power & Electric Co.
		Cambridge Electric Light Company
		Commonweatlh Electric Company

NERC region or subregion	NERC defined control area	Company name—NOPR jurisdictional
		New England Power Company
		New England Electric Transmission Corporation
		Massachusetts Electric Company
		Narragansett Electric Company
		Eastern Edison Company
		Montaup Electric Company
		Newport Electric Corporation
		Blackstone Valley Electric Company
		Concord Electric Company
		Exeter & Hampton Electric Company
NPCC NEPOOL.	NEPEX	Bangor Hydro-Electric Company
		Fitchburg Gas & Electric Light Company
		Unitil Power Corporation
NPCC-NYPP	NYPP	Central Hudson Gas & Electric Corporation
		Consolidated Edison Co-NY Inc
		Long Island Lighting Company
		Long Sault, Inc.
		New York State Electric & Gas Corporation
		Niagara Mohawk Power Corporation
		Orange & Rockland Utilities, Inc.
		Rochester Gas & Electric Corporation
		Rockland Electric Co.
SERC-FLA	FPC	Florida Power Corporation
	FPL	Florida Power & Light Company
	TEC	Tampa Electric Company
SERC Southern.	SCC	Alabama Power Company
		Georgia Power Company
		Gulf Power Company
		Mississippi Power Company
		Savannah Electric & Power Company
SERC-TVA		
SERC-VACAR.	CPL-E	Carolina Power & Light Company
	CPL-W.	
	DUK	Duke Power Company
		Nanthala Power & Light Co.
		Lockhart Power Company
	SCE&G	South Carolina Electric & Gas Company
	VAP	Virginia Power Company
		Old Dominion Electric Coop.
	YAD	Yadkin, Inc.
SPP	CLEC	Central Louisiana Electric Company, Inc.
	EDE	Empire District Electric Company
	EES	Arkansas Power & Light Company
		Gulf States Utilities Company
		Louisiana Power & Light Company
		Mississippi Power & Light Company
		New Orleans Public Service Inc.
	KCPL	Kansas City Power & Light Company
	MPS	UtiliCorp United Inc.
	WEI.	
	OGE	Oklahoma Gas & Electric Company
	PSO/SWEPCO.	Public Service Company of Oklahoma
		Southwestern Electric Power Company
		Rayburn Country Electric Coop
	SJLP	St. Joseph Light & Power Company
	SPS	Southwestern Public Service Company
	WRI	Western Resources Inc.
		Kansas Gas & Electric Company
WSCC	AZPS	Arizona Public Service Company
	EPE	El Paso Electric Company
	IPC	Idaho Power Company

NERC region or subregion	NERC defined control area	Company name—NOPR jurisdictional
	MPC	Montana Power Company
	NEVP	Nevada Power Company
	PACE	PacifiCorp
	PACW.	
	PG&E	Pacific Gas & Electric Company
	PGE	Portland General Electric Company
	PSC	Public Service Company of Colorado
		Cheyenne Light Fuel & Power Co.
	PNM	Public Service Company of New Mexico
		Texas-New Mexico Power Co.
	PSPL	Puget Sound Power & Light Company
	SCE	Southern California Edison Company
	SDGE	San Diego Gas & Electric Company
	SPPC	Sierra Pacific Power Company
	TEP	Tucson Electric Power Company
	WWPC	Washington Water Power Company
Public Utilities operating within Control Areas listed in Attachment B		
		Black Hills Corp. (in WAUM)
		Maine Public Service Co. (in MAR)
		MDU Resources Group Inc. (in WAWT)
		Minnkota Power Cooperative (in WAWT)
		Northwestern Public Service Co. (in WAUM)
		Northwestern Wisconsin Electric Co. (in DPC)
		Peoples Electric Cooperative (in WEFA) [1]
		Sho-Me Power Cooperative (in AECI)
Totals—P.U. Control Areas 76		
Total U.S. Control Areas—140; NOPR Public Utilities—166		

[1] People's status as a public utility is being contested at the Commission.

[Attachment B]

Electric Utilities, not subject to the Commission's jurisdiction under sections 205 and 206 of the FPA, that operate Control Areas.

NERC defined		Electric utility name
NERC region	Control area	
ECAR	BREC	Big Rivers Electric Corporation
	EKPC	East Kentucky Power Cooperative
	HEC	Hoosier Energy Rural Electric Cooperative
ERCOT	COA	City of Austin, TX
	CPS	City Public Service Board—San Antonio, TX
	LCRA	Lower Colorado River Authority
	PUB	Public Utility Board—City of Brownsville, TX
	STEC	South Texas & Medina Electric Cooperatives
	TMPP	Texas Municipal Power Pool
	HLP	Houston Lighting & Power Company
	TUE	Texas Utilities Electric Company
MAAC		
MAIN	CWL	Columbia Water & Light Department—Columbia, MO
	CWLP	CityCity Water, Light & Power—Springfield, MO
	SIPC	Southern Illinois Power Cooperative
MAPP	DPC	Dairyland Power Cooperative
	LES	Lincoln Electric System—Lincoln, NE
	MPW	Muscatine Power & Water, Muscatine, IA
	NPPD	Nebraska Public Power District
	OPPD	Omaha Public Power District
	SMMPA	Southern Minnesota Municipal Power Agency
	UPA	United Power Association
	WAWT	Western Area Power Administration—Watertown, SD
		Upper Missouri—East

NERC defubed		Electric utility name
NERC region	Control area	
NPCC		
SERC	AEC	Alabama Electric Cooperative
	FMP	Florida Municipal Power Pool
	FTP	City of Ft. Pierce, FL
	GVL	Gainesville Regional Utilities, FL
	HST	City of Homestead, FL
	JEA	Jacksonville Electric Authority, FL
	KEY	City of Key West, FL
	KIS	City of Kissimmee, FL
	LWU	City of Lake Worth, FL
	NSB	City of New Smyrna Beach, FL
	RC	Reedy Creek Improvment District
	SC	South Carolina Public Service Authority
	SEC	Seminole Electric Cooperative
	SME	South Mississippi Electric Power Association
	STC	City of St. Cloud, FL
	STK	City of Starke, FL
	TAL	City of Tallahassee, FL
	TVA	Tennessee Valley Authority
	VB	City of Vero Beach, FL
SPP	AECI	Associated Electric Cooperative, Inc.
	BPU	Board of Public Utilities—Kansas City, KS
	CAJN	Cajun Electric Power Cooperative
	CIPL	City of Independence Power & Light, MO
	GRDA	Grand River Dam Authority
	LAFA	City of Lafayette, LA
	LEPA	Louisiana Energy & Power Authority
	SECI	Sunflower Electric Cooperative
	SPA	Southwestern Power Administration
	WEFA	Western Farmers Electric Coop.
WSCC	BPA	Bonneville Power Administration
	CHPD	Chelan County Public Utility District #1, WA
	DOPD	Douglas County Public Utility District #1, WA
	GCPD	Grant County Public Utility District #2, WA
	IID	Imperial Irrigation District
	LDWP	Los Angeles Department of Water and Power
	PASA	City of Pasadena, CA
	SCL	Seattle City Light, WA
	SRP	Salt River Project, AZ
	TCL	Tacoma City Light, WA
	WACM	Western Area Power Administration—Colorado Missouri
	WALC	Western Area Power Administration—Lower Colorado
	WAUM	Western Area Power Administration—Upper Missouri-West
Total—64		
Total U.S.—140		
Canadian	MAR	Maritime Area (Maine Public Service Co. operates within this control area)

O

91-929 (40)